Understanding the Elements of the Periodic Table™

LEAD

Kristi Lew

82   207

Pb

rosen publishing's
rosen
central

New York

*For all the men and women dedicated to keeping
lead products out of the hands and mouths of children*

Published in 2009 by The Rosen Publishing Group, Inc.
29 East 21st Street, New York, NY 10010

Copyright © 2009 by The Rosen Publishing Group, Inc.

First Edition

Library of Congress Cataloging-in-Publication Data

Lew, Kristi.
Lead / Kristi Lew.—1st ed.
    p. cm.—(Understanding the elements of the periodic table)
Includes bibliographical references and index.
ISBN-13: 978-1-4042-1779-9 (lib. bdg.)
1. Lead. 2. Periodic law—Tables. 3. Chemical elements. I. Title.
QD181.P3L49 2009
546'.688—dc22

                                        2007042140

Manufactured in Malaysia

On the cover: Lead's square on the periodic table of elements. Inset: The atomic structure of lead.

Contents

Introduction

Forensic scientists are looking into a 180-year-old mystery. What killed Ludwig van Beethoven (1770–1827), the famous classical music composer? Could it have been his doctor? Forensic scientist Christian Reiter of the Medical University of Vienna in Austria believes that it is possible. Many people know that Beethoven started to become deaf in his late twenties. But they are often surprised to learn that, before his death in March 1827, Beethoven had endured a terrible illness for more than thirty years.

In the days after Beethoven's death, mourners came to pay their respects to the great composer. Not only did they view Beethoven's body, but some mourners also clipped a lock of his hair to take home as a keepsake. In 2005, scientists used these hairs to investigate the reason for Beethoven's consistently poor health. They tested some of the hair and from it found that Beethoven's blood contained almost 100 times more lead (chemical symbol: Pb) than normal. It turned out that Beethoven had a severe case of lead poisoning.

Then, in 2007, Reiter tested two more strands of Beethoven's hair. Because hair absorbs chemicals—including lead—from the bloodstream as it grows, Reiter was able to establish a day-by-day "chemical diary" of the last four months of the composer's life. Medical records show that during

The German composer Ludwig van Beethoven was a victim of lead poisoning and died in 1827.

this time, Beethoven suffered from painful swelling in his abdomen. Scientists believe that this was due to cirrhosis, a liver disease. Four times during Beethoven's last months of life, his doctor, Andreas Wawruch (1771–1842), drained the fluid from his abdomen by puncturing it with a needle. Reiter's research shows that the amount of lead in the composer's body went up after each of these punctures. Scientists are not exactly sure where the lead came from, but they theorize that it could be from lead-containing medicines used by Wawruch to clean or seal the wound. These medicines may have made Beethoven's lead poisoning even worse, hastening his death. So, it is possible that in trying to help Beethoven, Wawruch may have accidentally killed the great composer instead.

Lead is quite poisonous to humans. It is also a cumulative poison, which means that it gradually builds up in the body. Over time, lead in the body can damage the brain, liver, and kidneys. Unlike some other metals such as zinc (Zn), iron (Fe), calcium (Ca), and sodium (Na), which are required for a healthy body, lead has no useful biological function and should not be eaten or otherwise taken into the body.

Lead is not all bad, though. The metal is quite useful as a shield for X-rays and the radiation produced in nuclear reactors. It also does a

Almost 180 years after Beethoven's death, researchers tested a lock of his hair and found that it contained massive amounts of lead.

good job of absorbing sounds and vibrations. Lead combined with other elements into chemical compounds can be used in lead-acid storage batteries that run cars, in delicate crystal bowls and glasses, and in making rubber stronger and more weather-resistant. Consequently, as long as it stays out of the human body, lead can be a very useful metal to have around.

Chapter One
A Closer Look at Lead

Lead is an element that has been known since ancient times. Elements are substances that are made up of only one type of atom. A piece of pure lead is made up of many tiny lead atoms. Atoms are the building blocks of all matter, which is anything that has mass and takes up space. A desk, the floor, and water are all types of matter—and so are you.

History of Lead

More than 100 elements make up the Earth. Lead is one of them. In fact, it was one of the first elements ever mined. Some lead coins and statues more than 7,000 years old have been found in Egyptian tombs. Nearly 2,500 years ago, the ancient Romans used lead, too. They got most of their lead from mines in England and Spain. Ancient Romans used the metal to make pewter plates (pewter is

Lead is a soft, heavy, toxic metal. When it is freshly cut, lead is a shiny, bluish white. But when it is exposed to air, the metal quickly turns a dull gray.

Ancient Romans built a network of pipes to carry water from natural springs into their towns. Some of the pipes were made of lead, including this one in the Roman baths in Bath, England.

a mixture of lead and tin [Sn]) and a large underground system of lead pipes to carry water. Some of these pipes can still be found in Rome today.

A Dangerous Substance

At one time, decorative house paint with lead as an added ingredient was very popular in the United States. People said that white lead, a paint pigment, or coloring, had an unusual brightness. Housepainters also reported that lead paint covered a house much better than paint made without lead. Certain types of glass are made with a lead compound, too. This type of glass is called lead crystal. Lead crystal chandeliers sometimes contain as much as 30 percent lead mixed in with the molten glass. Adding lead to the glass makes it sparkle brilliantly.

But as far back as 400 BCE, the physician Hippocrates (460–377? BCE) suspected trouble when he reported a lead miner with a severe case of abdominal pain. Marcus Vitruvius Pollio (85–25? BCE), a Roman engineer, also noticed that men who worked in smelters, places where lead was heated up to purify it, looked unusually pale and sickly. Later, doctors discovered that housepainters who used lead paint often developed "wrist drop," a condition in which the painters lost all control over their wrist muscles.

The Alchemists

Alchemists used lead. It was the first and, therefore, oldest of the seven metals of alchemy. The other metals were gold (Au), silver (Ag), mercury (Hg), copper (Cu), iron, and tin. Alchemists tried very hard to find a way to turn "base" (low in value) metals such as lead into "noble" metals, like silver or gold. But because one element cannot be turned into another element under normal circumstances, the alchemists were unsuccessful in their attempts.

Despite the apparent problems of working with lead, it was not until studies, conducted during the twentieth century, confirmed these dangers to human health that many countries banned lead paint and many other uses of the metal.

Where Is Lead Found?

Lead makes up about 0.0013 percent of the Earth's crust. Even though that sounds like a small amount, lead is not considered a rare metal because it is easily mined and purified. Pure lead, or lead not combined with any other elements, can sometimes be found in nature—but not very often. That is because lead reacts easily with other elements. When two elements react with each other chemically, the result is called a chemical compound. In nature, lead is most often found chemically bonded to the nonmetallic element sulfur (S). Scientists call this chemical compound lead(II) sulfide (PbS). Lead(II) sulfide is sometimes called by its common name, galena. Galena is an ore of lead. A lead ore is a deposit of lead that can be mined easily enough to make it financially worthwhile. Other

Metallic lead is rare in nature. Instead, the metal is usually found chemically bonded to other elements. The ore mimetite [chemical formula: $Pb_5(AsO_4)_3Cl$] is an example of a lead compound.

lead ores include anglesite [lead(II) sulfate, $PbSO_4$], cerussite [lead(II) carbonate, $PbCO_3$], and minum (lead tetraoxide, Pb_3O_4). Worldwide, more than two million tons (1.8 million metric tons) of lead ore are mined each year, primarily in the United States, Australia, Mexico, Germany, and France.

Some Properties of Lead

Lead is just one of the 111 elements that scientists have confirmed so far. (Elements with atomic numbers 112 and higher have been reported but have not been fully verified.) An essential property of an element is that it cannot be broken down into simpler substances by ordinary chemical means, such as exposure to acid, electricity, light, or heat. However, elements do combine with other elements to form chemical compounds.

Each element has its own special properties. Lead's physical properties, for example, include the fact that the element is a soft solid at room temperature (68° Fahrenheit [20° Celsius]). The metal is also malleable and ductile. Malleable means that lead can be hammered, rolled, or otherwise bent into different shapes without breaking. Ductile is a similar property that means that lead can be drawn into wires. In fact, a sheet of lead as thick as a nickel coin is so soft that it can be bent easily by hand. Lead has these properties because the bonds between the individual lead atoms in a piece of lead are fairly weak. Because the atoms are somewhat free to move around, the metal can be deformed without breaking.

Compared to other metals, lead has a low melting point. Lead melts at 621.43°F (327.46°C). Gold, on the other hand, melts at 1947.52°F (1064.18°C) and silver at 1763.20°F (961.78°C). The reason that lead has a lower melting point has to do with the bonds between the lead atoms. Because those bonds are weak, the metal will melt at a lower temperature than metals with stronger bonds between their atoms. It does not take much heat to soften lead. In fact, a hunk of lead could be softened over a campfire and molded into all types of useful objects, including pots, pans, and water pipes.

State of matter, malleability, ductility, and melting point are physical properties. Lead has chemical properties, too. Chemical properties describe how lead interacts with other elements. For example, one of lead's most useful chemical properties is its resistance to corrosion. Corrosion is what happens when a metal reacts with chemicals in the environment and starts to change into different substances. For example, when iron corrodes, the iron changes to rust. Lead's resistance to corrosion does not come from the fact that it does not react with other elements—it does. When pure lead is exposed to oxygen (O_2) in the air, it quickly reacts with the oxygen to

Chemical properties describe how substances interact chemically. For example, two colorless liquids, lead(II) nitrate [$Pb(NO_3)_2$] and potassium iodide (KI), chemically react to form a bright yellow solid, lead(II) iodide (PbI_2).

form a layer of lead(II) oxide (PbO). This layer of lead(II) oxide then reacts with carbon dioxide in the air to form a tough film of lead(II) carbonate on the outside of the piece of lead. The lead(II) carbonate layer prevents air and moisture from getting to the pure lead underneath and prevents further corrosion.

Chapter Two
Atomic Lead

Chemists use the symbol Pb to represent lead. This chemical symbol comes from the word *plumbum*, the Latin name for lead. The system of pipes that carries water to the kitchens and bathrooms in buildings is called plumbing, and this word is also derived from *plumbum* because ancient Romans used lead to make pipes to carry their water.

Lead and the Periodic Table

All of the chemical elements that scientists have discovered so far are listed on the periodic table. (See the periodic table of elements on pages 40–41.) The periodic table is a chart that organizes all of the elements. The elements are arranged in rows by increasing atomic number. The rows of the periodic table are also called periods. Lead is in period 6. The columns of the periodic table are called groups (or families). Lead is in group 14 (or IVA), along with the elements carbon (C), silicon (Si), germanium (Ge), and tin.

Dmitry Mendeleyev (also spelled Dmitri Mendeleev, 1834–1907) was one of the first chemists to produce a periodic table. He published his periodic table in 1869. In his periodic table, Mendeleyev arranged the approximately sixty chemical elements known during his time into a chart in order of increasing atomic weight. (Atomic weight is the average

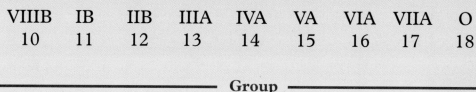

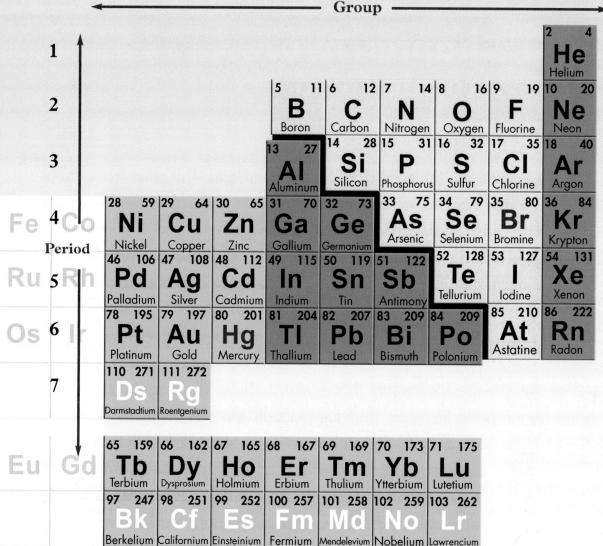

The chemical symbol for lead is Pb. On the modern periodic table, lead is found in period 6 and group 14 (or IVA).

weight of all the naturally occurring isotopes of a specific element, taking into account the proportions in which the isotopes occur. Isotopes are atoms of a specific element with different numbers of neutrons.) When Mendeleyev arranged the elements in this way, he noticed that many of their properties showed a periodic, or regular, pattern. Elements with similar chemical properties appeared in the same column or group.

Mendeleyev also saw that to keep the regular pattern of properties, he had to leave some blanks in his periodic table. He predicted that, eventually, other elements would be discovered that would fill in the blanks in his chart. He even predicted some of the properties of these unknown elements. Other scientists were doubtful. But soon, Mendeleyev would be proved

Henry Moseley found that an element's properties depended on its number of subatomic particles, not its weight. He rearranged Mendeleyev's periodic table, putting the elements in order by atomic number.

correct by the discoveries of gallium (Ga) in 1875, scandium (Sc) in 1879, and germanium (Ge) in 1886. Each of these new elements fit into one of the blanks in Mendeleyev's periodic table and exhibited properties that he had predicted.

As more elements were discovered, though, chemists began to find problems with arranging the elements by atomic weight. In 1913, Henry Moseley (1887–1915), a British physicist, discovered that what really determined an element's properties was its subatomic particles.

Atomic Structure

Atoms can be divided into three types of subatomic particles: protons, neutrons, and electrons. Protons and neutrons are found inside the nucleus, or center, of the atom. Electrons move around the outside of the nucleus. Protons are subatomic particles that have a positive charge. The number of protons determines the type of atom. All lead atoms, for example, have eighty-two protons. Any atom that does not have eighty-two protons is not a lead atom. The number of protons in an atom of an element is equal

Recycling Lead

Not all lead is mined. The metal is very easy to recycle, and it can be re-melted over and over again without changing its properties. Recycling lead saves energy and keeps the metal out of landfills, where it can pollute the environment. It requires 35 percent to 40 percent less energy to purify recycled lead than it takes to get pure lead from mined ore. About 70 percent of the lead used in the United States comes from recycled lead.

A lead atom's nucleus contains 82 positively charged protons. On average, the nucleus also contains 125 neutrons. Moving around the nucleus, 82 negatively charged electrons can be found in six energy levels.

to the element's atomic number on the periodic table. Lead has eighty-two protons and its atomic number is eighty-two. The periodic table is now arranged by atomic number—or the number of protons in an atom's nucleus—instead of by atomic weight.

Atoms are neutral. They have no charge. That means there must be an equal number of negative charges in an atom in order to balance the number of protons. Electrons are subatomic particles with negative charges. A neutral atom has the same number of electrons as it has protons. Electrons are found outside of the atom's nucleus in energy levels, or shells. The negative electrons are held around the atom by their attraction to the oppositely charged protons in the atom's nucleus. Because lead atoms have eighty-two protons, they also have eighty-two electrons, and these are arranged in six energy levels.

The third type of subatomic particle, the neutron, does not have a charge at all. Neutrons are neutral. But neutrons do have mass. So do protons. A proton has a mass of about 1.0 atomic mass unit (amu). A neutron is also about 1.0 amu. Electrons have a tiny mass—so tiny that they usually are not counted in the mass of an atom. Therefore, in order to find an element's mass, you must add its protons and neutrons.

On the periodic table, lead's atomic weight is listed as 207.2 amu. To determine the number of neutrons in an average lead atom, round the atomic weight to a whole number (207) and subtract the number of protons (82). Therefore, on average, lead atoms have 125 neutrons.

Isotopes

All atoms of the same element have the same number of protons. But they do not all have the same number of neutrons. Atoms of the same element that have a different number of neutrons are called isotopes. Scientists have identified more than forty lead isotopes. But only four of these isotopes occur in nature: lead-204, lead-206, lead-207, and lead-208. The number behind the element's name is the mass of that isotope. All lead atoms have eighty-two protons, but lead-204 has 122 neutrons, lead-206 has 124 neutrons, lead-207 has 125 neutrons, and lead-208 has 126 neutrons, making their masses different.

About 52 percent of the lead atoms found in nature are lead-208. Lead-207 makes up about 22 percent of lead atoms, and 24 percent are lead-206. Lead-204 is the least common. Only about 1.5 percent of lead atoms are lead-204. If the weights of these isotopes are averaged, taking into account which isotopes occur in nature the most, lead's atomic weight is 207.2 amu, the mass listed on many periodic tables. However, on our periodic table, the atomic weight of Pb has been rounded to three digits, 207.

Chemical Bonding and Lead

The electrons on an element's highest, or outermost, energy level are called valence electrons. When elements chemically react with one another, these valence electrons are lost, gained, or shared. When metals such as lead are involved in a chemical reaction, they often give up their valence electrons to nonmetals that take the electrons.

When metals give up their electrons, which are negatively charged, they become positively charged because they have more protons than electrons. Lead has four valence electrons. Sometimes, lead loses only two of its valance electrons. Other times, it loses all four. If a lead atom loses

Lead Snapshot

Chemical Symbol:	Pb
Classification:	Metal
Properties:	Bluish-white color; soft; malleable; ductile; corrosion-resistant
Discovered By:	Was known since ancient times
Atomic Number:	82
Atomic Weight:	207.2 atomic mass units (amu)
Protons:	82
Electrons:	82
Neutrons:	122, 124, 125, or 126
State of Matter at 68°F (20°C):	Solid
Melting Point:	621.43°F (327.46°C)
Boiling Point:	3180°F (1749°C)
Commonly Found:	Combined with the element sulfur in the lead ore galena

When a colorless solution of lead(II) nitrate [$Pb(NO_3)_2$] is added to sulfuric acid [H_2SO_4], a chemical reaction between the two liquids forms a white solid called lead(II) sulfate [$PbSO_4$].

two of its valence electrons, it has a charge of +2. If it loses all four, its charge will be +4. These charged particles are called ions. The electrons lost by lead atoms are gained by atoms of a nonmetal such as oxygen. An oxygen atom can gain two electrons to become a negative ion, with a charge of −2. The opposite charges of the positive lead ion and negative oxygen ion are attracted to each other. This attraction causes the ions to stick together, forming a chemical bond between them.

Because lead ions can have two different charges, chemists have come up with a way to distinguish between them. They do this by using a Roman numeral in the names of lead compounds. For example, PbO is lead(II) oxide. This compound contains a lead ion with a +2 charge. PbO_2, on the other hand, is lead(IV) oxide. There is another, older method of naming these two different forms of lead, too. This classical method can still be found in some books. In this older naming method, lead(II) is called plumbous and lead(IV), plumbic. So, lead(II) oxide would become plumbous oxide. Lead(IV) oxide would be named plumbic oxide. Scientists rarely use this method of naming today, though.

L ead was once used more than it is now, but there are still some important applications of the metal. Because lead can have harmful effects on living things, though, care needs to be taken to keep it out of the environment as much as possible.

Lead-Acid Storage Batteries

Most of the lead produced in the United States is used to make lead-acid storage batteries. These types of batteries are the primary source of electricity in today's cars. The battery is an electrochemical device. Electrochemical devices do not store electricity. They store chemicals that produce electricity through a series of chemical reactions. During these chemical reactions, the battery changes chemical energy into electrical energy. That electrical energy can be used to run a car's electrical devices.

In most automobile batteries, a reddish-brown substance called lead dioxide (PbO_2) covers the positive terminal of the battery. The positive terminal of a battery is called the cathode. Lead dioxide can also be called lead(IV) oxide. The negative terminal, or anode, of the battery is made up of a porous type of lead called sponge lead. Both of these substances are spread onto metal grids to make up positive and negative plates inside the battery.

Most of the lead that is produced in the United States today is used to make lead-acid storage batteries for cars.

These metal grids are most commonly made of a lead-antimony alloy. An alloy is a mixture of two or more elements, at least one of which is a metal. Alloys are often used because they have different properties from either of the elements they are made up of. In the battery's lead-antimony plates, for example, an alloy is used because lead by itself is very soft. The antimony in the alloy makes the grids harder and stronger than if pure lead were used.

The metal grids sit in a mixture of sulfuric acid and water. Sulfuric acid is a strong acid and is very corrosive. It can easily eat through clothes and cause severe injury to both the skin and eyes. However, lead is resistant to corrosion, even by a strong acid like sulfuric acid. The sulfuric acid is an electrolyte. An electrolyte is a substance that conducts electricity when it is dissolved in water or melted. It does this by allowing electrons or ions to move between the cathode and the anode in the battery. Electricity is the movement of charged particles such as electrons or ions.

When the car's battery is used—to start the car, or to power the lights or radio while the car is off, for example—the chemical composition of the metal plates and the sulfuric acid is changed. The sponge lead on the anode plates reacts with sulfuric acid and gradually turns into lead(II) sulfate. In doing this, the lead atoms release electrons and become lead

+2 ions. The electrons travel from the battery through wires to the car's starter or lights or radio, where they make the electrical devices operate. From there, the electrons continue through wires back to the cathode plates in the battery. At the cathode, the electrons combine with the lead +4 ions, changing them into lead +2 ions. These ions combine with sulfuric acid and form lead(II) sulfate. The acid strength weakens as the battery is discharged.

When the car engine runs, it drives the car's alternator. The alternator is an electrical generator, and it changes the direction of the electrical current through the battery. Reversing the direction of the current restores the chemical composition of the plates and the acid, recharging the battery.

Electronics

Another common use for lead is in lead soldering. There are four main ways to join materials together: mechanical joining (using screws, nuts, or bolts), adhesive joining (gluing), welding (melting metals together to join them), and soldering. In soldering, two solid metal parts are joined by melting a third metal on top of them, making a joint that fuses the two metals together.

An alloy of lead and tin was once a popular solder used to join parts on circuit boards. Today, lead solder has largely been replaced by other, less toxic metals.

Solders are classified by their melting temperature. Hard solders usually contain alloys of silver, copper, and zinc and require high temperatures to melt. Soft solders have lower melting points. Most soft solders were made of a lead-tin alloy. Lead solder is often used to make connections on circuit boards. Circuit boards can be found in many electronic devices, including computers and televisions. Because of concerns about the polluting effects of lead on the environment, lead is no longer used in most solders and electrical devices currently being manufactured.

Other Uses

Lead is a very dense metal. This characteristic means that, compared to other substances, lead has more mass per unit volume. Because of its high density, lead is very effective at absorbing radiation. Radiation is energy moving through space. Visible light, ultraviolet (UV) rays from the sun, X-rays, and radio waves are all examples of radiation. Some types of radiation, called ionizing radiation, can be harmful to people. Sometimes, when people have X-rays taken, a lead apron is used to shield body parts

Pencil Leads?

Despite the name, pencil leads have no elemental lead in them. Instead, they are made up of a non-toxic mixture of graphite (a form of the element carbon) and clay. When a graphite deposit was discovered in Borrowdale, England, in 1564, people thought the graphite was a type of lead. In fact, they called it black lead, or plumbago. This might explain why it is called pencil "lead" instead of pencil "graphite."

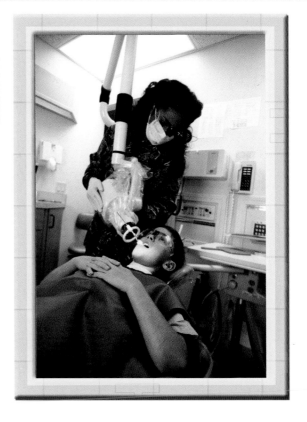

Lead aprons, such as this one used in a dental procedure, absorb radiation emitted by an X-ray machine, protecting body parts that do not need to be X-rayed.

that do not need to be X-rayed. This keeps those body parts from being exposed to X-ray radiation.

Lead bricks are sometimes used as a shield around nuclear reactors so that they can absorb any radiation that might accidentally leak out. The glass in television screens also contains lead to shield people from the radiation emitted by the television. Televisions and computer monitors contain about one-half pound (227 grams) of lead. People need to be very careful about how they dispose of these products.

Absorbing vibration and sound is another way that people take advantage of lead's relatively high density. For this reason, lead is sometimes used in the construction of hotels and modern office buildings where noise might be a problem. Because of the metal's high density, only a thin layer of lead is needed to block out sound.

Lead is also a good choice when a corrosion-resistant metal is needed. It is used to cover some electrical wires and cables to keep them from being damaged by moisture or chemicals in the environment. In addition, acid storage containers are sometimes lined with lead when an acid might eat through other materials.

Lead sheeting is lead that has been formed with rollers into large, flat sheets. These sheets have been used in the building industry as flashing. Flashing is sheet metal that has been shaped and applied around angles

Waterbirds can accidentally swallow lead fishing weights, causing the birds to die of lead poisoning. To protect them, sinkers are being made with less toxic metals.

of a roof or wall to prevent water from leaking into the house. Lead was used because it is tough yet flexible. Plus, it lasts for a long time and people liked the way it looks. However, due to our increased knowledge of the harmful effects of lead on the environment, lead flashing is being replaced by other metals such as copper.

Lead weights were once very popular in fishing, but because of environmental concerns, they largely have been phased out. However, lead weights can still be found as curtain weights, weights for some chemical laboratory equipment, and in sailboat keels to give the boats stability.

Chapter Four
Lead Compounds

Lead is used in its pure form, and it is also combined with other elements in chemical compounds. But because scientists have determined that lead is toxic to humans and that it lingers in the environment for a long time, lead compounds are not as popular as they used to be.

Leaded Gasoline

Gasoline sold in the United States once contained the additive tetraethyl lead [$Pb(CH_2CH_3)_4$]. Gasoline is a mixture of chemical compounds that contain hydrogen and carbon. These compounds are called hydrocarbons. In modern cars, a gasoline-air mixture is compressed before the spark plug ignites the fuel. But some hydrocarbons ignite before they are fully compressed. This occurrence causes the car to run roughly and to make clattering, or "knocking," sounds.

In 1922, Thomas Midgley (1889–1944), an American engineer, discovered that adding tetraethyl lead to gasoline caused the hydrocarbons to burn more slowly and smoothly, silencing the knock. To remove the lead from the car's cylinder, another chemical was added. This chemical reacted with the tetraethyl lead and released a vapor of lead(II) bromide ($PbBr_2$) into the environment in the car's exhaust.

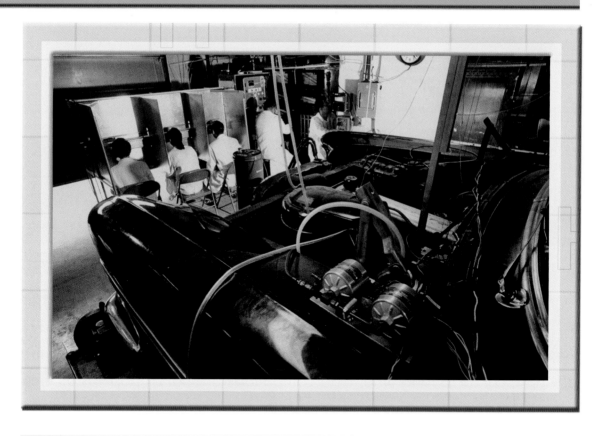

This car's exhaust fumes are being tested in 1960. Lead added to gasoline made cars run better, but it also polluted the environment. Leaded gasoline was banned in the United States in the 1970s.

There were two major problems with leaded gasoline. The lead in the car's exhaust was toxic to humans, and the lead in the gasoline clogged and destroyed catalytic converters. Catalytic converters are added to car engines to reduce the amounts of the polluting gases carbon monoxide (CO) and nitrogen oxides (NO and NO_2) released into the atmosphere.

Because of these problems, leaded gasoline was phased out in the United States in the 1970s. Since then, safer gasoline additives have been discovered. However, in 2007, seventeen countries around the world were still using leaded gasoline.

Because they tend to put objects—like paint chips—into their mouths, babies and young children are the most likely age group to be poisoned by lead in their environment.

Lead Paint

Many lead compounds are used as paint pigments, or colorings. Lead(II) chromate ($PbCrO_4$), also called crocoite, is used to make a bright yellow pigment called chrome yellow. Chrome yellow is still used as the pigment in road markings.

Lead(II) carbonate, or cerussite, was once used in white paint because it is a brilliant white color. In fact, lead carbonate is sometimes called white lead. Titanium dioxide (TiO_2) has now taken lead carbonate's place as the white pigment of choice, since it is non-toxic and much more stable than lead carbonate.

Lead(II) molybdate ($PbMoO_4$) is a reddish-orange pigment. So is trilead tetraoxide (Pb_3O_4), which is sometimes called red lead. Trilead tetraoxide was once used as a reddish-brown paint to protect outdoor furniture and other outdoor structures from corrosion, or rust.

Decorative lead paints have been banned in the United States because of health reasons. But like leaded gasoline, lead paint is still used in some countries, including China, India, and Malaysia. It may still be encountered in some older homes, too.

Adding the chemical compound lead(II) oxide to glass makes it very clear and heavy. Lead crystal is sometimes used to make glasses, bowls, and chandeliers that sparkle brilliantly.

Lead Crystal

Lead(II) oxide, sometimes called litharge, continues to be used to make lead crystal, a heavy type of glass. One way to make glass is to melt silica (SiO_2), soda (also known as sodium carbonate [Na_2CO_3]), and lime (CaO) together. To make lead crystal, lead(II) oxide takes the place of lime in the recipe. Leaded glass shines brilliantly. The added lead makes the glass clearer, heavier, and less likely to break when it is etched or carved.

This type of glass produces a characteristically high ringing sound when a moist finger is rubbed along the rim of the glass. In 1759, Benjamin Franklin (1706–1790) attended a concert in which a musician used this characteristic of lead crystal to play the "singing" glasses. The "singing" glasses were lead crystal goblets filled with differing amounts of water. Each glass emitted a different musical note depending on how much water it contained.

In 1761, Franklin invented a musical instrument that made similar sounds. But instead of having lots of wine glasses, his instrument had different-sized lead crystal bowls nested together and attached to an iron rod turned by a foot pedal. To play the instrument, the musician touched a damp finger to the spinning bowls. The bowls emitted the same haunting melodies as the wine glasses filled with water. Franklin painted

By touching a damp finger to the different-sized lead crystal bowls in Benjamin Franklin's glass armonica, musicians could make haunting melodies.

each bowl a different color so that musicians could tell which note they were playing. He called his invention the glass armonica. *Armonica* is the Italian word for "harmony."

Sadly, some armonica players began to fall ill. As a result, superstitions ran wild. Some people claimed that the haunting melodies brought forth evil spirits who made the musicians go mad. Others thought the music itself made the musicians ill, and many people refused to listen to armonica music. Scientists now believe it is possible that lead poisoning was the real culprit behind the musicians' illnesses. Not only were the bowls made of lead crystal, but the paints used on them were lead-based as well. By the 1800s, the glass armonica was all but forgotten.

Although lead crystal glasses and bowls are quite pretty, they can cause problems if the lead in the glass leaches, or leaks, into the food or beverages stored in them. The amount of lead that leaches out of the glass depends on how much lead is in the crystal, how long the food is stored in the crystal container, and what type of food is being stored. Acidic foods and drinks—like tomatoes, fruit juices, soft drinks, and wine, for example—increase the amount of lead released from the crystal. The longer these foods are left stored in the crystal, the more lead they may contain, too.

Vulcanization of Rubber

Lead crystal glasses and bowls are not the only objects in which lead(II) oxide is used. It is also employed in the vulcanization of rubber. Pure rubber, or latex, which comes from the sap of certain tropical trees, is almost useless for making car tires. The rubber is usable at room temperature, but it turns brittle and cracks if it gets too cold. Moreover, in hot weather it turns into a sticky mess. In an effort to remedy these problems, Charles Goodyear (1800–1860) performed many experiments with rubber in the mid-nineteenth century. One day, Goodyear accidentally dropped some

In this illustration, Charles Goodyear shows a group of fellow scientists his vulcanized rubber. Vulcanizing rubber makes it weather-resistant and more usable.

rubber mixed with sulfur onto a hot stove. The process, which later became known as vulcanization, resulted in the waterproof, weather-resistant rubber that manufacturers use in products today. High heat and an oxide, such as lead(II) oxide or zinc oxide, is also needed to make the vulcanization process work. The process was named for Vulcan, the Roman god of fire.

Other Uses for Lead Compounds

Lead(II) arsenate [$Pb_3(AsO_4)_2$] was once used as an insecticide because, just as lead is toxic to humans, it is poisonous to insects, too. However, it has largely been replaced with other, more environmentally friendly and less toxic insecticides.

When lead(II) nitrate [$Pb(NO_3)_2$] is heated, it emits a crackling noise. So does lead tetraoxide. These lead compounds are sometimes used in fireworks to produce this sound, but there are other chemicals, such as bismuth trioxide (Bi_2O_3) and bismuth subcarbonate [$(BiO)_2CO_3$], that can take their place and are less harmful to the environment.

Chapter Five
Lead and You

When the human body absorbs lead, the lead goes into the bloodstream. From there, it can circulate and enter vital organs such as the brain, liver, and kidneys. Any lead that is not eliminated by the body in a few weeks is stored in the bones and teeth, where it can stay for decades. During pregnancy, serious illness, or when a bone breaks, lead stored in bone can be re-released into the body's bloodstream. This can be particularly dangerous during pregnancy because lead can cross the barrier of the placenta and get into the baby's bloodstream. The nervous system of an unborn baby is particularly sensitive to even small amounts of lead.

Health Effects of Lead

For many years, people made products from lead for use in and around their homes. At the time, they did not know that lead was toxic and could cause health problems such as learning disabilities, behavioral problems, seizures, and even death, especially in children. If high lead levels in children are not detected early, their brains and nervous systems can be damaged. That is because lead is a neurotoxin, or a substance that is poisonous to nerve tissue.

The Downfall of the Roman Empire

According to some theories, widespread lead poisoning may have caused the downfall of the Roman Empire. Historical records have indicated that Roman food and wine may have been heavily contaminated with lead. In 1983, scientists tested the skeletal remains of Romans killed in the Mount Vesuvius volcanic eruption (79 BCE). They found that the bones contained abnormally high levels of lead. This excessive amount makes the scientists believe that it is very possible that the ancient Romans suffered from chronic lead poisoning.

Scientists have found abnormally high lead levels in the remains of ancient Romans buried in the volcanic eruption of Mount Vesuvius.

Because children are growing quickly and their bodies absorb more lead than adults, unborn babies and children under the age of six are especially at risk. Young children also have a tendency to put objects in their mouths, and these objects may be covered in lead dust. Lead is a cumulative poison, which means that it builds up in the body over time. A high level of lead in the bloodstream affects adults, too. In adults, it can lead to nerve disorders, high blood pressure, and muscle and joint pain. It can also cause problems with memory, concentration, digestion, and reproduction.

Toys that contain lead-based paint can be a hazard to small children. Workers planning to destroy toys contaminated with lead must protect themselves from breathing in lead dust as well.

Unfortunately, lead poisoning is not always easy to detect. Some people may have flu-like symptoms or just feel tired and weak. Lead poisoning can cause anemia, a blood disorder that results in a low red blood cell count. Because healthy red blood cells carry oxygen from the lungs to all of the other cells in the body, a low red blood cell count means that the body's cells are starved for oxygen. This condition causes symptoms such as fatigue and dizziness. Other symptoms of lead poisoning can include loss of appetite, nausea, vomiting, headaches, and abdominal cramps. Exposure to even low levels of lead over a long period of time can affect the nervous system and the brain, causing memory loss and depression. But some people have no symptoms at all until the level of lead in their blood becomes very high.

Many homes built before 1978 have lead-based paint in them. As long as the paint in the home is in good shape, it is probably not a problem. But paint that has begun to peel, flake, or get chalky needs to be properly removed. Lead can get into the body by breathing or swallowing lead dust, or by eating soil or paint chips that contain lead. So, proper removal techniques are very important to make sure that the lead dust is minimized. In 1978, the federal government banned the use of lead-based paints in homes.

Some homes built before 1978 may still have lead-based paint in them. When they remove the paint, workers must be careful to protect themselves and the environment from lead dust.

In 1992, the U.S. government passed another law requiring people selling a house or renting out an apartment built before 1978 to tell the buyers or renters about any lead-based paint hazards that the dwelling may have. This law is called the Residential Lead-Based Paint Hazard Reduction Act of 1992. It is also known as Title X (Title Ten). The same law requires businesses that renovate or remodel homes to provide the people who hire them with information about how to protect themselves from lead dust that might be disturbed by the remodeling.

Scientists have now determined that the health effects caused by lead are due to the metal bonding to sulfur in proteins inside cells. Because atoms of lead are bigger than those of the metals to which sulfur normally bonds, like zinc, lead twists the proteins out of shape. A protein that is incorrectly shaped does not function the way it should.

Combating Lead Poisoning

In the body, lead is stored in fat, soft tissue, and bones. Scientists have found that eating a low-fat diet that is rich in iron and calcium may help prevent the absorption of lead by the body. It is also a good idea to avoid eating canned food imported from other countries, where lead solder may

Students at a California school created posters to help educate the public about the dangers of lead poisoning and the proper ways to dispose of lead-containing products.

be used in sealing the can. And do not store food or fruit juices in cans. Because lead can still be found in the soil, even after leaded gasoline and lead-based paints have been banned for years, it is best to wash all vegetables and fruits before eating them. In addition, remove and discard the outer leaves of vegetables such as lettuces and cabbages. And peel root vegetables such as carrots and onions before eating them.

When gardening, do not plant food crops in soil around busy roadways, where lead-contaminated soil may still linger. In urban areas where traffic is concentrated, building raised garden beds and using fresh gardening soil bought at a gardening center can reduce the amount of lead in the garden's produce.

Finally, if a home was built before 1985, it is possible that lead soldering was used to join the water pipes in the home. Therefore, avoid using hot tap water for making drinks or baby formula or for cooking, since hot water can leach more lead out of the solder than cold water does.

With these precautions, as well as some thought as to how objects that contain lead are disposed of, people can still safely use the metal to help run their cars. They can also still use it to protect them from radiation and to absorb noise and vibration without adding more lead to the environment and harming the health of themselves and others.

The Periodic Table of Elements

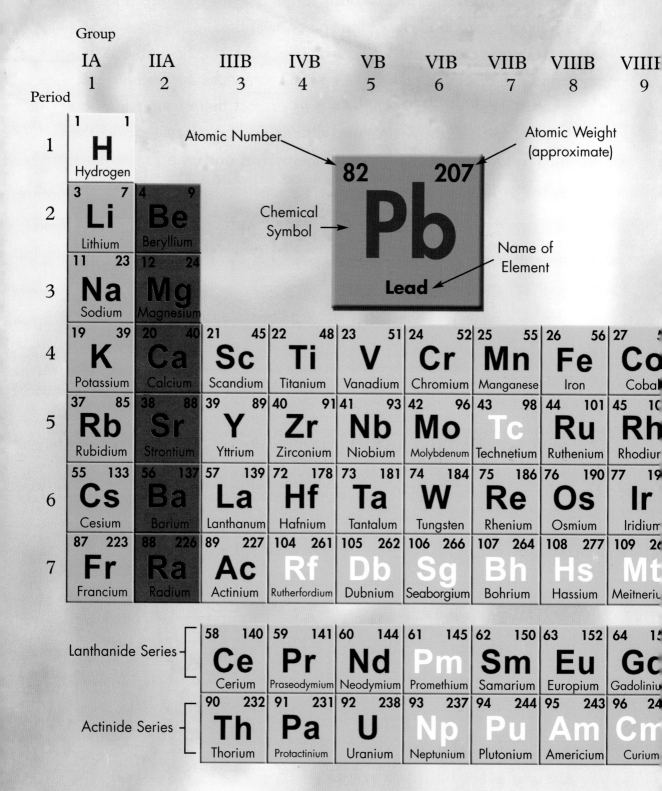

Group

IA	IIA	IIIB	IVB	VB	VIB	VIIB	VIIIB	VIIIB
1	2	3	4	5	6	7	8	9

Period

Atomic Number → 82   207 ← Atomic Weight (approximate)

Chemical Symbol → **Pb**

Lead ← Name of Element

Period									
1	1 1 **H** Hydrogen								
2	3 7 **Li** Lithium	4 9 **Be** Beryllium							
3	11 23 **Na** Sodium	12 24 **Mg** Magnesium							
4	19 39 **K** Potassium	20 40 **Ca** Calcium	21 45 **Sc** Scandium	22 48 **Ti** Titanium	23 51 **V** Vanadium	24 52 **Cr** Chromium	25 55 **Mn** Manganese	26 56 **Fe** Iron	27 **Co** Coba
5	37 85 **Rb** Rubidium	38 88 **Sr** Strontium	39 89 **Y** Yttrium	40 91 **Zr** Zirconium	41 93 **Nb** Niobium	42 96 **Mo** Molybdenum	43 98 **Tc** Technetium	44 101 **Ru** Ruthenium	45 10 **Rh** Rhodiur
6	55 133 **Cs** Cesium	56 137 **Ba** Barium	57 139 **La** Lanthanum	72 178 **Hf** Hafnium	73 181 **Ta** Tantalum	74 184 **W** Tungsten	75 186 **Re** Rhenium	76 190 **Os** Osmium	77 19 **Ir** Iridium
7	87 223 **Fr** Francium	88 226 **Ra** Radium	89 227 **Ac** Actinium	104 261 **Rf** Rutherfordium	105 262 **Db** Dubnium	106 266 **Sg** Seaborgium	107 264 **Bh** Bohrium	108 277 **Hs** Hassium	109 26 **Mt** Meitneriu

Lanthanide Series

58 140 **Ce** Cerium	59 141 **Pr** Praseodymium	60 144 **Nd** Neodymium	61 145 **Pm** Promethium	62 150 **Sm** Samarium	63 152 **Eu** Europium	64 15 **Gd** Gadoliniu

Actinide Series

90 232 **Th** Thorium	91 231 **Pa** Protactinium	92 238 **U** Uranium	93 237 **Np** Neptunium	94 244 **Pu** Plutonium	95 243 **Am** Americium	96 24 **Cm** Curium

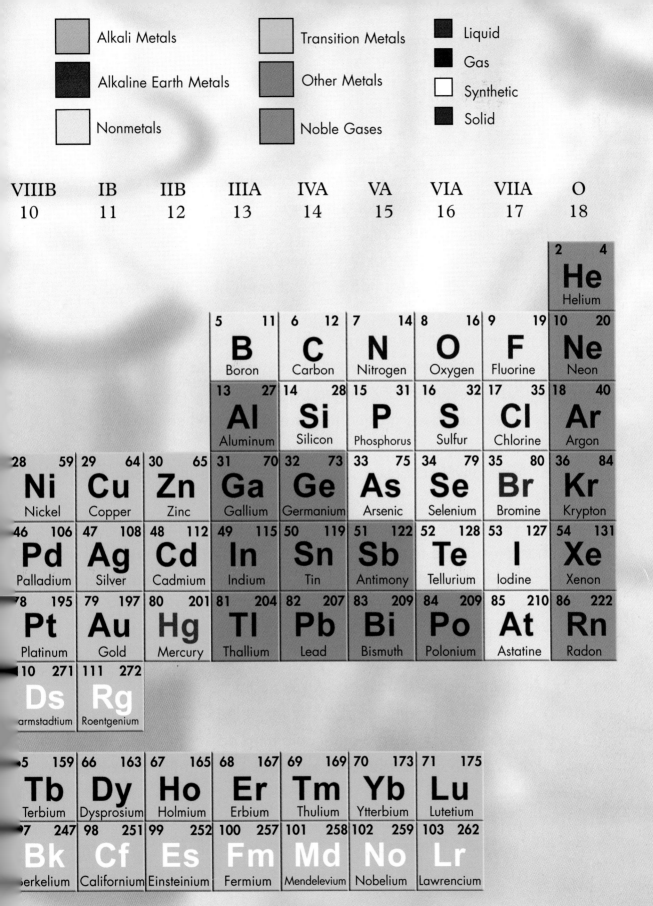

Legend

- Alkali Metals
- Alkaline Earth Metals
- Nonmetals
- Transition Metals
- Other Metals
- Noble Gases
- Liquid
- Gas
- Synthetic
- Solid

VIIIB 10	IB 11	IIB 12	IIIA 13	IVA 14	VA 15	VIA 16	VIIA 17	O 18
								2 4 **He** Helium
			5 11 **B** Boron	6 12 **C** Carbon	7 14 **N** Nitrogen	8 16 **O** Oxygen	9 19 **F** Fluorine	10 20 **Ne** Neon
			13 27 **Al** Aluminum	14 28 **Si** Silicon	15 31 **P** Phosphorus	16 32 **S** Sulfur	17 35 **Cl** Chlorine	18 40 **Ar** Argon
28 59 **Ni** Nickel	29 64 **Cu** Copper	30 65 **Zn** Zinc	31 70 **Ga** Gallium	32 73 **Ge** Germanium	33 75 **As** Arsenic	34 79 **Se** Selenium	35 80 **Br** Bromine	36 84 **Kr** Krypton
46 106 **Pd** Palladium	47 108 **Ag** Silver	48 112 **Cd** Cadmium	49 115 **In** Indium	50 119 **Sn** Tin	51 122 **Sb** Antimony	52 128 **Te** Tellurium	53 127 **I** Iodine	54 131 **Xe** Xenon
78 195 **Pt** Platinum	79 197 **Au** Gold	80 201 **Hg** Mercury	81 204 **Tl** Thallium	82 207 **Pb** Lead	83 209 **Bi** Bismuth	84 209 **Po** Polonium	85 210 **At** Astatine	86 222 **Rn** Radon
110 271 **Ds** Darmstadtium	111 272 **Rg** Roentgenium							

65 159 **Tb** Terbium	66 163 **Dy** Dysprosium	67 165 **Ho** Holmium	68 167 **Er** Erbium	69 169 **Tm** Thulium	70 173 **Yb** Ytterbium	71 175 **Lu** Lutetium
97 247 **Bk** Berkelium	98 251 **Cf** Californium	99 252 **Es** Einsteinium	100 257 **Fm** Fermium	101 258 **Md** Mendelevium	102 259 **No** Nobelium	103 262 **Lr** Lawrencium

Glossary

alloy A mixture of two or more elements, at least one of which is a metal.

anode The negative terminal of a battery.

atom The smallest unit of matter.

cathode The positive terminal of a battery.

compound A substance formed by the chemical bonding of two or more elements.

density A measure of the mass per unit volume of a substance.

ductile A physical property of a metal making it easy to form it into wires.

electricity The movement of charged particles.

electrolyte A substance that conducts electricity when dissolved or melted.

electron A negatively charged subatomic particle found outside the nucleus.

element A substance made up of only one type of atom.

ion A charged particle formed by the loss or gain of electrons.

isotope A form of an element with a particular number of neutrons and mass.

malleable A physical property of a metal allowing it to be bent without breaking.

matter Anything that has mass and takes up space.

neutron A neutral subatomic particle found in the nucleus of an atom.

nucleus The tiny center of an atom, containing nearly all of its mass.

proton Positively charged subatomic particles found in the nucleus of an atom.

radiation Energy moving through space.

valence electrons Electrons in the highest energy level of an atom.

Centers for Disease Control and Prevention (CDC)
Childhood Lead Poisoning Prevention Branch
4770 Buford Highway (Mail stop F-40)
Atlanta, GA 30341
(770) 488-3300
Web site: http://www.cdc.gov/nceh/lead/default.htm
Frequently asked questions (FAQs) about lead poisoning are on the
    CDC's Web site.

Lead Development Association International (LDAI)
17a Welbeck Way
London W1G 9YJ
+44 (0)20 7499 8422
Web site: http://www.ldaint.org
The LDAI produces publications and holds meetings and conferences all
    over the world to promote the safe use of lead and its compounds.

National Institute of Environmental Health Sciences (NIEHS)
P.O. Box 12233
Research Triangle Park, NC 27709
(919) 541-0395
Web site: http://niehs.nih.gov
NIEHS is the organization within the National Institutes of Heath (NIH)
    that studies the effects of the environment on human health. It explains
    numerous environmental health topics, including lead exposure, on
    http://kids.niehs.nih.gov/home.htm.

National Safety Council
1121 Spring Lake Drive
Itasca, IL 60143-3201
(630) 285-1121
Web site: http://www.nsc.org
Lead dust test kits can be purchased through the National Safety
    Council. These kits determine if there is a lead dust problem in your
    home. They include instructions, tools to take two dust samples, and
    a postage-paid mailer to send these samples to a U.S. Environmental
    Protection Agency–certified lab for analysis.

U.S. Environmental Protection Agency (EPA)
Ariel Rios Building
1200 Pennsylvania Avenue, NW
Washington, DC 20460
(202) 272-0167
Web site: http://www.epa.gov
The EPA teaches people about the environment and what they can do to
    protect it. You can download or order its print publications at
    http://www.epa.gov/teachers/order-publications.htm.

Web Sites

Due to the changing nature of Internet links, Rosen Publishing has
developed an online list of Web sites related to the subject of this book.
This site is updated regularly. Please use this link to access the list:

http://www.rosenlinks.com/uept/lead

For Further Reading

Baldwin, Carol. *Material Matters: Metals*. Chicago, IL: Heinemann Library, 2005.

Miller, Ron. *The Elements: What You Really Want to Know*. Minneapolis, MN: Twenty-First Century Books, 2006.

Spilsbury, Louise A., and Richard Spilsbury. *Elements and Compounds*. Chicago, IL: Heinemann Library, 2007.

Stwertka, Albert. *A Guide to the Elements*. 2nd ed. New York, NY: Oxford University Press, 2002.

Symes, R. F. *Rocks and Minerals*. New York, NY: DK Publishing, Inc., 2004.

Thomson, Ruth. *Recycling and Re-Using Metal*. North Mankato, MN: Smart Apple Media, 2006.

Tocci, Salvatore. *Lead*. Danbury, CT: Children's Press, 2005.

Watt, Susan. *Lead*. New York, NY: Benchmark Books/Marshall Cavendish, 2002.

Whyman, Kathryn. *Metals and the Environment*. North Mankato, MN: Stargazer Books, 2004.

Zronik, John Paul. *Metals*. New York, NY: Crabtree Publishing Co., 2004.

Bibliography

Blanc, Paul. *How Everyday Products Make People Sick: Toxins at Home and in the Workplace.* Los Angeles, CA: University of California Press, 2007.

Centers for Disease Control and Prevention. "Lead." Retrieved September 21, 2007 (http://www.cdc.gov/lead).

The Franklin Institute. "A Magical Touch . . . of Harmony." Retrieved September 21, 2007 (http://fi.edu/qa99/attic9/index.html).

Health Canada. "Lead Crystalware and Your Health." Retrieved September 21, 2007 (http://www.hc-sc.gc.ca/iyh-vsv/prod/crystal_e.html).

Michael, Andrea. "Lead and Food." University of Minnesota Extension. Retrieved September 21, 2007 (http://www.extension.umn.edu/info-u/nutrition/BJ652.html).

Ornes, Stephen. "Beethoven Dead from Lead?" ScienceNOW Daily News, *Science* Magazine, August 28, 2007. Retrieved September 21, 2007 (http://sciencenow.sciencemag.org/cgi/content/full/2007/828/1?etoc).

Sullivan, Kevin R. "12-Volt Lead Acid Battery Basics." Retrieved September 21, 2007 (http://www.autoshop101.com/trainmodules/batteries/101.html).

U.S. Environmental Protection Agency. "Lead in Paint, Dust, and Soil." August 2, 2007. Retrieved Sept. 21, 2007 (http://www.epa.gov/lead/pubs/leadinfo.htm#facts).

Wilford, John N. "A Clue to the Decline of Rome." *New York Times*, May 31, 1983. Retrieved September 21, 2007 (http://query.nytimes.com/gst/fullpage.html?sec=health&res=9D04E3DE163BF932A05756C0A965948260&n=Top%2fReference%2fTimes%20Topics%2fSubjects%2fL%2fLead).

About the Author

Kristi Lew is a professional K–12 educational writer who has degrees in biochemistry and genetics. A former high school science teacher, Lew specializes in writing textbooks, magazine articles, and nonfiction books about science, health, and the environment for students and teachers.

Photo Credits